QUALITY RULES in PACKAGING

REVISED AMERICAN EDITION

John Sharp

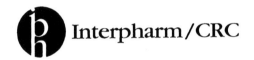

Boca Raton   London   New York   Washington, D.C.

This book contains information obtained from authentic and highly regarded sources. Reprinted material is quoted with permission, and sources are indicated. A wide variety of references are listed. Reasonable efforts have been made to publish reliable data and information, but the author and the publisher cannot assume responsibility for the validity of all materials or for the consequences of their use.

Neither this book nor any part may be reproduced or transmitted in any form or by any means, electronic or mechanical, including photocopying, microfilming, and recording, or by any information storage or retrieval system, without prior permission in writing from the publisher.

The consent of CRC Press LLC does not extend to copying for general distribution, for promotion, for creating new works, or for resale. Specific permission must be obtained in writing from CRC Press LLC for such copying.

Direct all inquiries to CRC Press LLC, 2000 N.W. Corporate Blvd., Boca Raton, Florida 33431.

Trademark Notice: Product or corporate names may be trademarks or registered trademarks, and are used only for identification and explanation, without intent to infringe.

Visit the CRC Press Web site at www.crcpress.com

© 2002 by John Sharp
Interpharm is an imprint of CRC Press

No claim to original U.S. Government works
International Standard Book Number 1-57491-132-5
Printed in the United States of America 1 2 3 4 5 6 7 8 9 0
Printed on acid-free paper

CONTENTS

	Page
Preliminary Note for Training Managers and Instructors	iv
1. Good Manufacturing Practice (GMP)	1
2. Making and Packaging Drug Products	7
3. GMP and the Law—The FDA	10
4. The Two Main Stages of Packaging	15
5. The Purpose of Packaging	17
6. The Packaging Operation	21
7. Printed Packaging Materials	25
8. On-line Checking	31
9. Summary of GMP Requirements in Packaging	33
10. The Importance of Packaging	40

Preliminary Note for Training Managers and Instructors

This Short Guide had its origins in the many operator training courses in basic GMP that I have presented at a large number of drug products manufacturing facilities, large and small, in a wide range of locations. Although I usually provide photocopied handouts, I continue to be aware of a need for a simple booklet covering the major content of these courses, that is, for an easy-to-read outline that the trainees can keep as a reminder, a refresher, and a reference. The official GMP Regulations and Guidelines are generally too lengthy, detailed, and user-unfriendly for such purposes. Hence the "Quality Rules" series of booklets, which, in addition to outlining the fundamentals of GMP, places these requirements in the context of a brief discussion of the background to drug product manufacture and its regulatory control. The first booklet in the series was *Quality Rules—A Short Guide to GMP.* It has been supplied (in its original form) and used successfully in all Continents (except, probably, Antarctica), and has recently appeared as a revised *American* edition.

This present booklet is concerned with GMP in the packaging of drug and similar products. However, so that it may stand alone, it does repeat, in modified form, some of the initial basic material that appeared in the first booklet.

It is not intended that this booklet should be regarded as a substitute for a properly organized and well-presented GMP training program, but as a useful adjunct to it.

John Sharp
September 2001

1. GOOD MANUFACTURING PRACTICE (GMP)

This booklet is about Packaging. More specifically, it is about Good Manufacturing Practice (GMP) in the Packaging of Drug Products (or Pharmaceuticals). It is important to remember that these products are used in the treatment of people who have some form of illness or disease, or who have other problems or injuries that need care. If they are the right QUALITY, properly made, **and properly packaged,** our products can do a lot of good, even save lives. If they are badly made, are poor quality, or have been badly packaged, they can fail to have the intended effect, be hazardous to health, or even cause death. You'd better believe it—people have died as a result of BMP ("bad manufacturing practice").

So, what **is** Good Manufacturing Practice?

It has been defined as:

"The part of Quality Assurance which is aimed at ensuring that products are consistently manufactured to a quality appropriate to their intended use."

Making sure, that is ASSURING, that products are the right quality is very important in all kinds of manufacturing industry. With the sort of products we make, where lives could be at stake, it is absolutely vital. It truly could be a matter of life and death.

QUALITY ASSURANCE involves many activities, from the original Research, Design, and Development stages, right through to Manufacture and Packaging, and onwards through Storage, Transport, and Distribution. If we get it wrong at **any** of these stages, the consequences for the Quality of our products, and for the safety and well–being of our consumers (or patients) could be very serious, harmful, or even fatal.

GMP is the day-to-day part of Quality Assurance. Quite simply, GMP is about all the things we have to do, and the care we have to take in our daily work, to ensure the QUALITY of our products. That is what it is all about—QUALITY. So, there are two reasons why this booklet is called *Quality Rules*. Sure, it contains some rules about Quality. But it also means that Quality "Rules," in the sense that Quality is tops, or the best, just as some guys might say "The White Sox Rule," or "Broncos Rule," and so on.

Nowhere is GMP more important than in Packaging. All the good and careful work that has gone before can be ruined by faulty packaging, with possibly very serious or harmful results.

So, what do we mean by "Quality"?

"Quality" is one of those words that can mean a lot of different things. Sometimes the word is used to mean "excellence" or "goodness." On the other hand, some people say that a thing is the right QUALITY when it is tested and found to meet (or comply with) a specification.

These uses of the word "QUALITY" are not wrong. It is just that there **are** different possible meanings. When we talk of QUALITY we mean something that is more simple, and yet has wider implications. We say:

QUALITY MEANS FITNESS FOR ITS INTENDED PURPOSE.

Any product that we supply is fit for its purpose only when:

- IT IS THE RIGHT PRODUCT.

- IT IS THE RIGHT STRENGTH.

- IT IS FREE FROM CONTAMINATION.

- IT HAS IN NO WAY DETERIORATED, DECAYED, OR BROKEN DOWN.

- IT IS IN THE RIGHT CONTAINER.

- IT IS CORRECTLY LABELED.

- IT IS PROPERLY SEALED IN ITS CONTAINER AND PROTECTED AGAINST DAMAGE AND CONTAMINATION.

That is, in a nutshell, **when it is fit to be used in treatment or diagnosis with the confidence that it will have the desired effects and not harm or damage the consumer or user, in any way, through faults in manufacture.**

Before reading any further, take another look back at the list (above) of the factors that make a product fit for its intended purpose, and notice what a big part packaging and labeling have to play.

Why is GMP so important?

Drug Products (or Pharmaceuticals) are different from most other manufactured products. When people obtain most of the other things they want or need, they can check on the Quality of those products before they buy. Afterwards, if the goods are found to be faulty, they can in many cases take them back for a replacement or refund. Patients taking drug products have very little chance of detecting if anything is wrong. Yet if something **is** wrong, it could be very dangerous. People tend to accept the quality of drug products very much on trust. They trust the doctor who prescribes them, the pharmacist who dispenses them, and the doctor or nurse who administers them. Ultimately, they **all** trust those who manufacture and package them. And that means us!

This places a big responsibility on all of us, for this is a trust we must not betray.

'But,' you might think, 'what about all the testing that is carried out on the products we make and package? Won't all the lab testing be able to detect and reject any faulty products?' Well, NO, it won't—and it is important to remember that we cannot rely just on testing to do that.

We always have to remember that:

- We can only carry out tests on samples, because the sort of tests we carry out destroy the product—and that leaves a lot of the batch untested

and also

- We cannot test for everything that might have gone wrong.

There is one other factor that makes GMP so important. With many other products, a few wrong or defective items may not do a lot of harm, apart from irritating a customer and perhaps losing some future sales. BUT, a wrong, defective, or incorrectly labeled drug product can seriously damage or even kill a patient. With drug products and other pharmaceuticals, even a very small proportion of defective, contaminated, or mislabeled items can be very dangerous indeed. Yet, the final consumer (or patient) has very little chance of detecting that anything is wrong until it may be too late.

So, we can sum up the REASONS FOR GMP as a combination of:

1. The poor chance that a patient has of detecting that anything is wrong.

2. The weakness of Product Testing, because

    a. We can only test samples

   and

    b. We cannot test for everything.

3. The dangers to patients of even only a small number of defective or wrongly labeled items in a batch (and it is very difficult to detect a small number of defectives).

That, then, is what GMP is, and the reasons for it. It is really all about the care and attention necessary to ensure that we get things right, and keep them right, from the start and all along the line. And, never forget, Packaging comes towards the very end of the line. As we said, mistakes in packaging can ruin all the good work that has gone before, with possibly very dangerous consequences.

Those who have read the first booklet in this series *(Quality Rules—A Short Guide to Drug Products GMP)* will know about:

THE TEN BASIC RULES OF GMP

1. Be sure you have the correct written instructions before any job is started.

2. Always follow those instructions EXACTLY with no "cutting corners" (ASK if you don't understand).

3. Ensure that the correct Materials are being used.

4. Ensure that the correct Equipment is being used, and that it is CLEAN.

5. Prevent Contamination and Mix-up.

6. Always guard against labeling errors.

7. Always work accurately and precisely.

8. Keep things (including yourself!) clean and tidy.

9. Always be on the lookout for mistakes, errors, and bad practice, and report them immediately ("covering up" for yourself or another guy could cost lives!).

10. Make clear, accurate records of what has been done, and the checks carried out.

Nowhere are these Ten Basic Rules more important than in Packaging.

Before we talk a little more about these things, it will be useful to know something of the background to drug products in general, and about how the Law controls the way they are made.

2. MAKING AND PACKAGING DRUG PRODUCTS

What **are** drug products?

Well, we all know more or less what they are, but since these are the sort of products that we are making and packaging, it is best to be absolutely clear about just what the words "drug products" really mean.

Drug products come in a variety of different forms, for example, liquids, powders, creams, ointments, tablets, capsules, pills, eye drops, lotions, injections, and so on. In fact, the Code of Federal Regulations of the US Food and Drug Administration (21 CFR) states:

> *Drug product* means a finished dosage form, for example, tablet, capsule, solution etc., that contains an active drug ingredient generally, but not necessarily in association with inactive ingredients. . . .

But it is not just what they look like. It is what they are *used for* that makes these products so important.

DRUG PRODUCTS are things that are administered to, or taken by, patients in order to:

- TREAT a disease or illness.

- PREVENT a disease or illness.

- ALLEVIATE (that is, EASE) the symptoms of a disease or condition (for example, pain, inflammation, or irritation).

- DIAGNOSE a disease or condition.

- MODIFY normal bodily functions.

As we have already mentioned, other words that mean more or less the same as "drug products" are "finished pharmaceuticals." Another is "medicines."

Drug products (or medicines) have been used for thousands of years. For much of this time, they were mainly made from plant (and sometimes animal) materials. Their use was often tied up with various forms of magic and witchcraft (In some parts of the world it still is).

Some of the herbs used by ancient people unquestionably had useful effects and a few of them are still in use today. However, we now know that many of the substances that were once used had no real value, and some of them would even have been harmful. If people got better, in many cases it was because they would have gotten better anyway, perhaps helped a little by the mysterious powers of "faith."

Although the story of medicine goes back thousands of years, the age of modern scientific medicine (that is, the age of medicines that really do "work") started only in the early part of the Twentieth Century. In fact, it was around the year 1910 that saw the first beginnings of the development and use of a wide range of synthetic chemical substances to treat disease in a way that was far more effective than ever before. The benefits of this modern era of Medicine to the quality and quantity of human life have been enormous. But like everything else, there are two sides to the coin. When making modern drug products, we just cannot afford to get it wrong. We must work with constant care and attention to QUALITY. We must follow the Basic Rules of GMP.

THE PHARMACEUTICAL INDUSTRY

Until the early part of the twentieth century, drug products were largely manufactured by individual chemists, druggists, or apothecaries working in their own small laboratories. These activities later became the work of the local dispensing pharmacist "making up" prescriptions, one at a time—liquid mixtures, powders, hand-made pills, tablets, and so on. However, as the modern age of medicine developed, pharmaceutical factories were established to provide all these things in quantity. Since around the 1940s, the Pharmaceutical Industry has really taken off, in the United States, in Europe, and worldwide, in response to the need (and demand) for the modern drug products that have proved so successful in treating disease.

Many millions (billions even) of prescription items are dispensed each year throughout the world. Most of these items are made by the Pharmaceutical Industry (particularly the US Pharmaceutical Industry). The present-day Pharmacist in the Drug Store usually makes only a very small proportion, if any, of the prescriptions he or she dispenses.

The majority of prescription drug products are also the results of the Research and Development activities of the Pharmaceutical Industry, and the beneficial effects on the health and well-being of the people have been remarkable. As compared with the beginning of the twentieth century, many more babies survive infancy, fewer children die of what were once "killer" diseases, people live longer and more comfortable lives, and a number of major diseases have been conquered. In the discovery and development of powerful new drug products, the US Pharmaceutical Industry has been, and is, a world leader.

So, modern drug products can, and have, done an enormous amount of good. But we must remember that if they are not made and packaged properly, they can be very dangerous. It is not surprising, therefore, that working in such an important industry, which has such an impact on people's health, requires a special code of practice. It is also not surprising that there are special US Regulations about the manufacture and supply of medicines.

3. GMP AND THE LAW—THE FDA

Back in those earlier days that we mentioned in the last section, that is, before the early part of the twentieth century, when drug products were relatively simple and made only on a very small scale, there was not much regulatory control of drug products manufacture. Most of the products available in those days, if they didn't do much good, generally could not do much harm. With the great changes in the Pharmaceutical Industry, which started in the early twentieth century, and which really took off after 1940, came the need for large-scale (and often very expensive) pharmaceutical facilities to manufacture large quantities of the powerful and highly effective drug products that were being developed. From the regulatory angle, it became a whole new ball game. Much stricter control was necessary to ensure that the drug products available to the people were safe to take, had the desired and intended good effects, and were properly made. Many countries throughout the world now have laws and regulations designed to give this assurance. The United States is in the forefront of this global activity. It was one of the first, if not *the* first, countries in the world to introduce detailed drug product regulations, and its regulatory control system remains one of the most comprehensive and most efficiently administered in the world.

In the United States, the government body responsible is the Food and Drug Administration (FDA). Before any new drug product may legally be made available, it is necessary to make a New Drug Application (NDA) to the FDA, and have it approved. The NDA must provide detailed evidence that the new drug is both safe and effective, and that it will be properly packaged. But it goes further than that. The FDA wants also to know that in day-to-day production, the product is made and packaged properly, and to the right Quality standards. That is, they want to be sure that manufacturers are working in accordance with the principles of Good Manufacturing Practice. They publish the **Current Good Manufacturing Practice**

Regulations (21 CFR Parts 210 and 211) often called "the **cGMPs.**" But rules and regulations are not much good unless there is some way of checking that they are obeyed. So the FDA has set up a body of experts, whose job it is to go check.

These experts are the FDA Investigators who visit drug product manufacturing sites to check out their manufacturing, packaging, and storage facilities, their manufacturing, packaging, and control operations, their equipment, laboratories, personnel, documentation, and records. They look for violations of the cGMPs. They are authorized to enter premises, inspect operations, and review documents and records. If they are not happy about what they see or hear, they can make recommendations to the Administration, which could result in the company's not being allowed to stay in business as a manufacturer of drug products. These FDA investigators also visit manufacturing sites outside our country. If they are not satisfied with what they see, the FDA can refuse to allow the product to be imported into the United States.

Much of the FDA's **cGMPs** (21 CFR Parts 210 and 211) apply generally to Packaging. They also contain a Subpart (Subpart G) that refers specifically to "**Packaging and Labeling Control.**" So, you see, Packaging is vitally important not just from the viewpoint of the health of the consumer (and the health of the Company). It is also important from the angle of complying with the regulations.

To give you an insight into what these regulations are about, here are some extracts from that Subpart G of the cGMPs:

Subpart G—Packaging and Labeling Control

§211.122 Materials examination and usage criteria.

(a) There shall be written procedures describing in sufficient detail the receipt, identification, storage, handling, sampling, examination, and/or testing of labeling and packaging materials; such written procedures shall be followed. . . .

(b) Any labeling or packaging materials meeting appropriate written specifications may be approved and released for use. Any labeling or packaging materials that do not meet such specifications shall be rejected to prevent their use in operations for which they are unsuitable.

(c) Records shall be maintained for each shipment received of each different labeling and packaging material indicating receipt, examination or testing, and whether accepted or rejected.

(d) Labels and other labeling materials for each different drug product, strength, dosage form, or quantity of contents shall be stored separately with suitable identification. Access to the storage area shall be limited to authorized personnel.

(e) Obsolete and outdated labels, labeling, and other packaging materials shall be destroyed.

(f) Use of gang-printed labeling for different drug products, or different strengths or net contents of the same drug product, is prohibited unless the labeling from gang-printed sheets is adequately differentiated by size, shape, or color.

(g) If cut labeling is used, packaging and labeling operations shall include one of the following special control procedures:

(1) Dedication of labeling and packaging lines to each different strength of each different drug product;

(2) Use of appropriate electronic or electromechanical equipment to conduct 100-percent examination for correct labeling . . . ; or

(3) Use of visual inspection to conduct a 100-percent examination for correct labeling . . . Such examination shall be performed by one person and independently verified by a second person

(h) Printing devices on, or associated with, manufacturing lines used to imprint . . . upon the drug product unit label or case shall be monitored to assure that all imprinting conforms to the print specified in the batch production record.

§211.125 Labeling issuance.

(a) Strict control shall be exercised over labeling issued for use in drug product labeling operations.

(b) Labeling materials issued for a batch shall be carefully examined for identity and conformity to the labeling specified in the master or batch production records.

(c) Procedures shall be used to reconcile the quantities of labeling issued, used, and returned, and shall require evaluation of discrepancies found . . . Such discrepancies shall be investigated. . . .

(d) All excess labeling bearing lot or control numbers shall be destroyed.

(e) Returned labeling shall be maintained and stored in a manner to prevent mixups and provide proper identification.

(f) Procedures shall be written describing . . . the control procedures employed for the issuance of labeling; such written procedures shall be followed.

§211.130 Packaging and labeling operations.

There shall be written procedures designed to assure that correct labels, labeling, and packaging materials are used for drug products; such written procedures shall be followed. These procedures shall incorporate the following features:

(a) Prevention of mixups and cross-contamination by physical or spatial separation from operations on other drug products.

(b) Identification and handling of filled drug product containers that are set aside and held in unlabeled condition for future labeling operations to preclude mislabeling of individual containers, lots, or portions of lots.

Identification need not be applied to each individual container but shall be sufficient to determine name, strength, quantity of contents, and lot or control number of each container.

(c) Identification of the drug product with a lot or control number that permits determination of the history of the manufacture and control of the batch.

(d) Examination of packaging and labeling materials for suitability and correctness before packaging operations, and documentation of such examination in the batch production record.

(e) Inspection of the packaging and labeling facilities immediately before use to assure that all drug products have been removed from previous operations. Inspection shall also be made to assure that packaging and labeling materials not suitable for subsequent operations have been removed. Results of inspection shall be documented in the batch production records.

§211.134 Drug product inspection.

(a) Packaged and labeled products shall be examined during finishing operations to provide assurance that containers and packages in the lot have the correct label.

(b) A representative sample of units shall be collected at the completion of finishing operations and shall be visually examined for correct labeling.

(c) Results of these examinations shall be recorded in the batch production or control records.

This has been a shortened version of the part ("Subpart G") of the cGMP Regulations that covers Packaging and Labeling Control. Like all legal documents, it does not make easy reading, but these are some of the regulations that all pharmaceutical manufacturers have to comply with.

4. THE TWO MAIN STAGES OF PACKAGING

Talking in a very general sort of way, we can say that there are two main stages in all (or at least most) packaging operations. These are:

1. PRIMARY PACKAGING—that is, the stage where the product (the liquid, the tablets, the cream, or whatever) is filled into the container (bottle, vial, tube, and so on) with which it will be in direct contact, and where that container is sealed with a lid, cap, or bung of some sort. Filling tablets or capsules into bubble or strip packs is also a primary packaging operation, because the tablets or capsules are in direct contact with the films or foils used to form the strip or bubble packs. The packaging materials used in primary packaging are called PRIMARY PACKAGING MATERIALS, or sometimes CONTACT PACKAGING MATERIALS. We can call the result of a primary packaging operation a PRIMARY PACK.

2. SECONDARY PACKAGING—that is, when the primary pack is placed in a carton, a box, or a tray, which in turn may then be film-wrapped, shrink-wrapped or foil-wrapped. Sometimes these secondary packs are even further wrapped or packed together (usually by tens or by dozens) using film, shrink-wrapping, or larger cartons or boxes ("outers"). All the materials used in secondary packaging are called SECONDARY PACKAGING MATERIALS, or sometimes NONCONTACT PACKAGING MATERIALS.

When modern, high-speed, automatic (or semiautomatic) packaging lines are used, secondary packaging often follows straight on from primary packaging, without a break, on the same packaging line. The big difference is that, whereas primary packaging materials come into direct contact with the product, secondary packaging materials do not.

Other important sorts of packaging material are PRINTED PACKAGING MATERIALS. Printed packaging materials may be either primary (for example, printed tubes, printed ampules, or printed plastic bottles) or secondary (for example, printed cartons or printed outers). Printed labels and any printed leaflets that may be included with the products may also be considered printed secondary packaging materials.

In terms of the quality and safety of our products, the careful control and use of printed packaging materials, including labels, is the most important single factor in packaging operations.

5. THE PURPOSE OF PACKAGING

What is the purpose of a package? In other words, what is packaging **for**?

There are at least two very simple, obvious reasons for putting a product in a package:

1. To **contain** a convenient (or useful) quantity of the product (We would not want to supply our products "loose," or by the pocketful or handful, would we?).

and

2. To **protect** the product during storage, transport, and distribution (a paper bag, for example, would not be good enough).

But there are a number of other good reasons. We could summarize the main functions of a package, including the two obvious ones, as follows:

- To **hold**, or contain, a defined quantity of the product.

- To **protect** the product from:
    - Damage,
    - Contamination,
  or
    - Deterioration.

- To **identify** the product (what it is, its batch number, who manufactured it, and so on).

- To indicate required storage conditions, expiry date, or shelf life.

- To provide **other information,** for example:
    - Directions for use, dosage,
    - Warning of any hazards in use,
    - Information on side effects, and so on.

- To present the product in a form that is **easy to use.**

- If the product is a medicine, to help in ensuring that patients take their medicines as, when, and how they should.

You will notice that one of the important things a package does is to **give information.** This is not just a good idea. Much of this information (product identity, batch number, manufacturer's name, expiry date, storage conditions, and so on) is also required by Law.

We also listed above **protection** of the product as one of the important purposes of packaging. This raises the question, "Protection from what?"

Our products can be harmed (and as a consequence lose activity, fail to work, or even become dangerous) by:

- **Mechanical damage**—Shaking, jarring, dropping, or other impact can break up tablets or capsules, or damage a device, or break or crack a container, allowing the product to leak out, or contamination to enter.

- **Heat**—A number of products will break down or deteriorate at raised temperatures. Although protection against heat is mainly a matter of storage, the package can help.

- **Light**—A number of products are sensitive to light, which can cause breakdown and/or discoloration. Opaque, or colored glass, containers can help protect against this.

- **Humidity**—Moisture can severely damage products, so the packaging must provide protection against the product getting damp.

- **The package itself** can also harm the product inside it, if it is a wrong or badly designed pack. Obviously, the product must not be able to leak or escape from the package. Some materials (for example, some plastics) can absorb substances from the product, or allow them to pass through. There is also the possibility that some of the materials used to make containers can release substances (and what is more, *dangerous* or *poisonous* substances) into the product. This applies particularly to some plastics, but some inferior grades of glass can release chemicals or glass flakes into liquid products. Also, some materials used to make containers can react chemically with some products.

So, any material that will:

- Allow the product to leak or seep through it,

- Release substances into the product,

or

- React with the product

must not be used to make containers that will be in direct contact with the product.

'Well, OK,' you might say. 'But those of us working on the packaging line have nothing to do with deciding what sort of containers should be used.' Sure—these are the problems that have to be sorted out by the people responsible for the original design of the pack. They are responsible for ensuring that the package they specify is **right** for the product and that it will protect it and not harm it in any way. **But it is also very important that the people doing the packaging make absolutely sure that the specified packaging materials are, in fact, used.**

So you see the vital importance of making sure that our products are packaged and labeled correctly. If they are not, products could become damaged, spoiled, or contaminated, or false information could be given about what they are, their strength, and how they should be used and stored.

The GOLDEN RULE about packaging is that we must make absolutely certain that the RIGHT PRODUCT is packaged into the RIGHT, CLEAN CONTAINER that is CORRECTLY LABELED, and also, that any other printed materials (like cartons and leaflets) are the correct ones for the job.

Perhaps that all sounds easy. Well, it is not! Over the years, the biggest number of recalls of drug products has been due to errors in packaging. Every effort must be made to prevent packaging errors, especially labeling errors.

6. THE PACKAGING OPERATION

Before we talk about the actual activity of packaging, let us say a little more about that **GOLDEN RULE.**

We need to ensure that:

- The product we are packaging is not only the RIGHT PRODUCT, but also that it is the RIGHT BATCH.

- The product is filled, or put, into the RIGHT PACK, all the components of which (that is, all the containers, caps, labels, cartons, and so on) are correct, clean, and just as specified by the Packaging Instructions.

- The containers are CORRECTLY LABELED, and are marked with the correct BATCH NUMBER, and, as necessary, such details as the correct manufacturing and expiry dates.

- The pack contains the correct amount of product—that is, the correct number, weight, or volume.

- The pack is properly capped or otherwise sealed.

We need to be absolutely sure about all these things. It is also important that the pack looks good. Cosmetic defects may not necessarily be a hazard to the consumer, or patient, but dirty or scruffy packs, crooked labels, smudged printing, and things like that are not good for a company's reputation or "image"— or for the confidence that consumers have in the product.

FILLING METHODS

The way a product is filled into its primary container depends on the nature of the product. Except in very small-scale operations (where traditional manual filling may be used), filling is usually performed automatically or semiautomatically by machines.

There is a whole range of automatic equipment that can be used to fill the required amount (number, weight, or volume) of product into a primary container. Tablets or capsules can be counted out and filled by counters that are either electronic or mechanical.

Blister, strip, and foil packs all involve "counting" the required number of tablets (or capsules) into the blister or strip itself. For example, in blister packaging, the roll of plastic film, as it is fed to the machine, does not at that stage have any blisters or "bubbles" in it. These are molded into the film in the machine, and the tablets (or capsules) are fed into them. This is a form of counting, since it is almost impossible to fill too many tablets (or capsules) into one blister. The only real possibility is that there might be some unfilled blisters. This should not happen on a well-designed, well-made, and well-maintained machine. However, just to be sure, there are usually visual and/or electronic checks as backup.

In any large-scale operation, liquids are usually filled from a bulk container on semiautomatic conveyor lines equipped with filling, capping, labeling, and cartoning machines. There are a number of different types of liquid filling machines. Two of the main types are Volumetric Fillers and Vacuum Fillers. A volumetric filler fills every bottle to the same volume (if it is properly adjusted, that is). So if a bottle is oversized, the fill level will look low. On the other hand, if the bottle is undersized, it will look overly full, or it will even overflow. In contrast, a vacuum filler fills to a constant level in the bottle. If the bottle is oversized, the consumer will receive an excess. If it is undersized, the consumer will receive a short volume. These examples show the importance of the standardization and Quality Control of packaging materials.

Ointments, creams, and pastes are usually filled into tubes or jars by volumetric fillers. Because of the thickness of the products, vacuum fillers are usually not suitable.

Usually, on a semiautomatic filling line, the filled containers are then conveyed further along for any additional packaging, for example the insertion of wadding for tablet packs, the application and tightening of caps, and the application of labels. Then may follow one or more stages of secondary packaging. As we have said, secondary packaging materials do **not** come into direct contact with the product, but nevertheless they can have some important effects on the Quality of our products—that is, on their fitness for their purpose.

REASONS FOR SECONDARY PACKAGING

There are a number of possible reasons for secondary packaging. These include:

- **Protection of the product**—The secondary packaging can provide an extra layer of protection. A carton, for example, can help prevent breaking of glass or crushing of tubes. It can also provide additional protection from light.

- **Product identification**—A carton or box can make available more space for printing the name, strength, quantity, batch number, and description of the product, plus its manufacturing and expiry date.

- **Instructions and information**—A carton or box provides more space to print instructions for use and other important information, like storage conditions. This might not be possible on a very small primary pack.

- **Incorporation of other items**—It is often difficult to include any extras if a primary pack is supplied just on its own. This can become possible when, say, a bottle is placed in a carton. These "extras" can be of different types. The most common are:
  1. **Dosing aids or measures**—things like droppers, spoons, applicators, and small plastic measures.
  2. **Leaflets** that give further information about the product and how it should be used.

- **Security**—A secondary pack can provide an extra level of security against tampering. It can make the primary pack more difficult to get at, or to get at without leaving obvious evidence of tampering. For example, flaps of cartons or boxes can be stuck down or taped over.

- **Ease of handling**—Orders from our customers often call for multiples of ten rather than single packs. Storage, handling, order-picking, dispatch, loading, and unloading are all made easier if the product is packed in cartons of ten.

- **Appearance**—Very often, a neat secondary pack improves and preserves the appearance of the pack. This may not seem as important as some of the other reasons for secondary packaging, but it can have important consequences for acceptability by the consumer, confidence in the product, and for the company's overall image.

So, you see, secondary packaging can be almost as important as primary packaging. In fact, although we may have talked about them as if they are two different operations, on modern packaging lines, they are usually both performed in one continuous operation. Such operations are much faster and more efficient than simple manual methods, but because such a lot of things are happening so quickly, it does mean that extra care is necessary to prevent packaging errors. That is particularly true where printed materials are concerned—but that deserves a section to itself.

7. PRINTED PACKAGING MATERIALS

Mistakes in labeling, that is, putting the wrong label on a container, or putting a labeled product in the wrong carton, and things like that, have been the cause of more recalls than all the other possible causes put together.

There is surely no need to say how dangerous—deadly even—wrongly labeled products can be. With drug products, it could mean patients taking the wrong product or the wrong dose. It is impossible to stress too much the need for constant care and attention to prevent mislabeling and similar errors from happening.

Usually there are two different stages of printing on packaging materials. We talk of "preprinted materials" when we mean labels, cartons, leaflets, and the like that are received already printed from a printer. If extra details, like batch numbers or expiry dates, are printed on these just before or during the packaging operation, this is usually called "overprinting."

Routinely used **preprinted** materials include:

- Labels.

- Printed cartons.

- Leaflets.

- Tubes, as used for creams and ointments (Some of these are preprinted, and some are plain and are labeled after printing).

- Printed ampules (Some ampules are not preprinted and are labeled after filling).

- Printed films, foils, and laminates, such as are used in blister, strip, and foil packages.

The sort of information that is usually preprinted includes:

- Product name.

- Dosage form ("Tablets," "Ointment," "Suspension," and so on).

- Strength ("10 mg per tablet," for example).

- Pack size or quantity (for example "50" or "100 mls").

- Manufacturer's name.

- Manufacturer's address.

- Information on use, safety, and storage.

- Any other information required by the regulations.

That is the sort of information that it is usually preprinted. It applies to all batches of a particular product, and so it can be printed on things like labels and cartons that will be used for a number of batches of a product, over a period of time. However, some of the information that needs to be printed will vary from batch to batch, and is usually overprinted in spaces left on preprinted material for that purpose. This can be done "off line" in an on-site print shop, just before the packaging operation, or "on line" during the packaging run.

Because it *does* vary from batch to batch, we call this "variable information." The variable information that is overprinted on labels and cartons includes:

- Product batch number,

- Date of manufacture,

and

- Expiry date.

Now, it is possible, by very careful planning (and with a very good, reliable service from a printing firm) to have the variable information preprinted. Most manufacturers, however use preprinted materials for each product and pack size, and then overprint the variable information.

Techniques used for overprinting range from simple hand stamping (for very small-scale operations), to more sophisticated methods like ink-jet printing or laser printing. Numbers or letters can also be impressed or embossed onto a tube or a carton.

TYPES OF LABELS

Suppliers of labels provide labels in two main formats. These are called:

- Cut labels

and

- Roll-feed labels.

Cut labels—Printers usually print this sort of label several at a time on large sheets, which are then cut, in the print works, into individual labels. They are then supplied as wrapped or boxed bundles. They are usually not self-adhesive. The printing of several labels at the same time on one large sheet of paper is called "gang printing." As you will know if you read the quotes from Subpart G of the cGMP Regulations (in Section 3 of this booklet), the use of labels for different products or for different strengths of the same product that have been gang printed together is prohibited, *unless* the labels are so different from each other, in terms of color, shape, or size, that they could not possibly be mixed up.

And this is the problem. Gang printing provides an opportunity for label mix-up at the printing firm, before even the labels have been received at the company using the labels in packaging. Cut labels are also sometimes called "loose labels."

Roll-feed labels are self-adhesive labels, supplied on a long rolled-up strip of backing paper.

Roll-feed labels are preferable, as they very much reduce the possibility of mix-ups, or of the odd loose label left lying around or in a labeling machine, which can so easily happen with cut labels. If cut labels are used to label products, then additional precautions are necessary. These are detailed in Subpart G of the cGMPs.

But, hear this—we **must not** allow ourselves to become overconfident. We should be aware that the use of roll-feed labels does not, alone, completely eliminate the possibility of mislabeling. But, they are a definite safety improvement over cut labels.

SECURITY IN THE USE OF LABELS AND OTHER PRINTED MATERIALS

There are a number of things that can be done to help prevent mistakes with labels and other printed materials. These include:

1. **Secure storage.** Printed packaging materials, after Quality Control approval, should be held in secure stores (preferably apart from other stores areas) that are locked when they are unsupervised, and where entrance is permitted to authorized staff only.

2. **Security on issue.** When required for use, printed packaging materials should be issued only to authorized persons, in specified quantities, and on presentation of formal requisition documents. THERE MUST BE NO CASUAL, UNCONTROLLED ISSUE OF LABELS and other printed materials.

3. **Security in transportation.** Special care needs to be taken to ensure security of printed materials during transport:
   - From printer to manufacturer.
   - From Goods-Receiving to stores.
   - From stores to packaging line.

   It is usually a good idea to use sealed boxes or "cages" for transportation of printed materials around the manufacturing facility.

4. **Following instructions.** All packaging operations should be performed *precisely* in accordance with the Batch Packaging Instructions. These will normally indicate the materials that are to be used. Before the operation starts, all the materials issued must be checked against the Batch Packaging Instructions to ensure that the correct materials and labels will be used.

5. **Counting and Reconciliation.** That is, issuing printed materials in known and specified quantities, and at the end of a batch packaging run, comparing the number of packs produced with the amounts of the materials issued, making allowances for any spoilages.

6. **Line clearance checks.** At the end of every packaging run, and before a new run starts, the entire line or location and surrounding areas should be checked to ensure that there are no products, containers, labels, cartons, or leaflets left over from a previous batch (see also Section 9).

7. **Code reading.** Labels and other printed materials can have printed on them codes that uniquely identify them. The most common is, of course, a bar code. Patterns of perforations have also been used. These codes can be read by electronic scanners, or Code Readers. The best way of doing this is to install Code Readers on the packaging line itself. A rather less satisfactory alternative is to perform the code reading separately, off line, before the materials are taken to the packaging line for use.

8. **CONSTANT VIGILANCE.** In spite of all mechanical and electronic aids, a major factor in the prevention of potentially fatal packaging errors is the constant care and attention of all packaging operators and checkers. It is impossible to overemphasize the importance of *people* working in packaging to the protection of patients and other consumers.

8. ON-LINE CHECKING

Preventing packaging errors requires the constant care and attention of everybody working on a packaging line—operators, checkers, supervisors—**everyone**. In spite of all the modern automatic and checking devices, packaging errors can still occur. It is up to all those who work in packaging to see that they do not. No one should think that it is some other guy's responsibility.

The sort of faults that can be detected by visual inspection, and that should be constantly watched for, include:

- Wrong labels or other printed materials.

- Damaged or torn labels.

- Badly applied labels (not straight, or not fully stuck onto the container).

- Other damaged or incorrect packaging materials.

- Poor print quality.

- Incorrect or poor quality overprinting.

- Wrong quantity of product filled.

- Contamination of product.

- Leaking or spilling of product.

Sure, there are several different mechanical and electronic devices that can be used for on-line checking—like missing label detectors, code readers or scanners, and automatic weight checkers. But it is important to remember that a checking or measuring device that is not working, or that is not working *accurately*, can be very dangerous. To just assume that code readers are working properly, so that "every thing must be OK," can have very hazardous consequences for patients.

All testing, checking, and measuring equipment should be checked before use to ensure that it is working properly and accurately.

9. SUMMARY OF GMP REQUIREMENTS IN PACKAGING

At the beginning of this booklet, we summarized the main issues in GMP as "The Ten Basic Rules of GMP." As an even briefer summary, we can say that GMP is about:

- **Following Instructions,**
- **Keeping Records,**
- **Avoiding Contamination,**
- **Preventing Mix-ups,**

and

- **Avoiding Labeling Errors.**

We have talked about some of these already, but there are still some more things that need to be said, and others that need to be reemphasized.

PACKAGING INSTRUCTIONS

All packaging operations will be carried out in accordance with some form of written procedure, or batch packaging instruction. This will be a copy of a carefully checked master document that will state, among other things, the name of the product to be packaged, its form and strength, the pack size and the quantities, and the sizes and types of all the materials to be used. It will also give instructions about performing the packaging operation and the equipment to be used, and about any overprinting, in-process controls, checks, and measurements that are required.

Now, it's fine having instructions, but the really important thing is that we must do what the instructions say. They must be followed EXACTLY.

The batch packaging instruction is often also used to form the RECORD of a specific packaging run, and further information must be added to it as the work proceeds.

The sort of information that is added to complete the record includes:

- The bulk batch number of the product being packaged.

- The dates/times of starting and finishing the packaging run.

- The signatures or initials of the person (or persons) who supplied the bulk product, and of those who checked and confirmed that they were correct.

- Batch identity number of all primary and all printed packaging materials used.

- Quantities of product and materials used.

- Quantity of packaged product produced.

- Results of any in-process controls and checks.

- Signatures, or initials, of those who carried out all significant operations or checks.

There are some Very Important Things to remember about making records (and that includes entering a check signature or initials). So we have:

RULES ABOUT MAKING RECORDS

1. **Always make records, or enter signatures or initials, when (or immediately after) an action or reading has been taken, or the check has been made.** Records should always be made as things happen, NOT at the end of the shift, day, or week. They are about current events, not past history!

2. **A person entering a second-check signature, or initials, is confirming that he or she actually saw what was done (for example, a weighing) and has personally checked that everything was correct** (product, material, batch, quantity, reading, or whatever). It is NOT good enough, and **it could be very dangerous**, to just "trust the other guy" to have got it right, and write in second-check initials/signatures at the end of the day. These, too, should be entered as things happen, when they have been SEEN to happen.

3. **Records should always be NEAT and CLEAR.** They do not have to be beautiful, but remember, others may need to read them in five or more years' time.

4. If you make a mistake when making an entry on a document, it is not a crime. **But if you do make a mistake, do not obliterate it or cover it up. Cross it out neatly (so that it can still be read), make the correction, sign/initial it, and add the date (with any explanation necessary).**

Always remember that if the instructions are followed exactly, and the checks required are carefully carried out, then that should ensure that only the correct product, the correct materials, and the correct equipment are used. But mistakes can happen, so workers on a packaging line must be on the lookout, all the time, for errors and mistakes.

So, that has covered the first two items ("Following Instructions" and "Keeping Records") in the list at the head of this section. What about Contamination, Mix-ups, and Labeling Errors? Well, we all need to take great care to make sure that those things do not happen. One very important special measure in Packaging is the Line Clearance Check.

LINE CLEARANCE CHECKS

Probably the biggest cause of contamination or mix-up is leftovers (product, labels, materials) from a previous batch.

Contamination from a previously packaged batch can occur because of:

- Ineffective cleaning of vessels, hoppers, or other containers of bulk products.

- Unclean, or switched, pipelines.

- Tablets, capsules, or pills left behind in hoppers, or in counting devices.

Similar or even worse hazards can occur due to labels or other materials left on the line or in label dispensers.

To guard against these, and before any packaging operation begins, checks should be made to ensure that the work area, line, and equipment are clean (really CLEAN) and completely clear of any product, product residues, materials, labels, or documents left over or not required for the new packaging run about to begin.

This should **not** be just a quick, casual look-see, but a very thorough and specific check, carried out in accordance with a written procedure, item by item. These checks should be carried out by persons authorized and instructed to do so. They should record on the written instructions (with signatures or initials) that each item has, in fact, been properly checked and found to be OK. The completed checklist should form part of the retained batch packaging documentation.

OTHER SOURCES OF CONTAMINATION

The previous batch packaged is not the only source of product contamination. Other possible causes include:

- **Dirty primary containers.** Primary containers should be supplied clean, or cleaned on line. Line operators should check that they are, in fact, clean.

- **Dust from dry products** (tablets, capsules, pills) packaged in the same area. This dust must be kept under control, or the product packaged in a separate segregated area.

- **Inadequate separation of packaging lines.** Different lines should be kept well apart, or have physical barriers between them. (It is quite amazing how tablets, capsules, or even entire packages can wander or jump from one line to the next!)

There is another possible source of contamination or mix-up. Can you think what it is?

It is the packaging operators themselves!

It is all too easy for serious mistakes to occur through carelessness by operators. Things like picking up labels, cartons, or products and putting them back in the wrong place, or carrying them about loosely or in pockets can be very dangerous. This sort of thing **must not happen**.

Also, people (all of us) shed many thousands of particles, skin flakes, and germs all the time. We DO NOT want these in our products. It is therefore important that the protective clothing supplied (including headwear) is worn properly. Products should not be handled unless you are instructed to, and then only if clean gloves are worn. Nobody, but nobody, should cough, sneeze, or splutter over products.

Cosmetics too can contaminate products by powdering, flaking, or peeling off. Jewelry and wristwatches can also be a source of contamination. The jewelry itself, or part of it, might fall off into a product, and the wearing of jewelry and watches tends to create warm, moist places where germs can flourish (feel under your watch on a warm day). It is generally considered that, in packaging areas (or at least where product is exposed), the only jewelry that may be worn is a plain wedding ring.

Here are some:

RULES ABOUT CLEANLINESS AND THE PREVENTION OF CONTAMINATION

1. Always maintain high standards of personal cleanliness—body, hair, and clothing, and always wash your hands after using the bathroom.

2. Always wear protective clothing properly, as instructed.

3. Never handle products directly by hand. If it is necessary to do so, wear the gloves provided.

4. If you have any infections (colds, coughs, stomach upsets, and things like that), or if you have any cuts or grazes, these must be reported to your supervisor, so that you can temporarily be given work where you can do no harm.

5. Eating, drinking, chewing, and smoking are not permitted in production, packaging, and stores areas, nor should food, drink, candy, smoking materials, or personal medicines be taken into these areas.

6. Always check that your work place and surroundings are clean and tidy.

7. If you have cleaning procedures to carry out, always follow the instructions EXACTLY.

8. Always check that equipment and containers have been properly cleaned and dried before you use them. It IS your responsibility.

9. Always work with care to avoid creating dust or spillages.

10. ALWAYS be on the alert for possible sources of contamination. If you see anything that looks as if it might cause contamination, report it immediately.

Finally:

Never assume that errors cannot possibly occur. That is another Golden Rule of GMP.

Constant care and attention are necessary to prevent errors and to detect them if they do occur. Nowhere is this more true than in Packaging.

10. THE IMPORTANCE OF PACKAGING

The objective of this booklet has been to help you to understand just how important Packaging is, and how important YOU are in ensuring that the products you help to package will be safe and effective in use.

Always remember that two of the essential things that a package does are PROTECT the product and give INFORMATION. If our products are not properly protected against damage, contamination, and deterioration, or if their labeling or other packaging gives false or incorrect information, then the final consumer could suffer very serious consequences. It is up to you to see that this does not happen.